POISONING PLANET EARTH

POLLUTION AND OTHER ENVIRONMENTAL HAZARDS

THE ENVIRONMENT: *OURS TO SAVE*

POISONING PLANET EARTH

POLLUTION AND OTHER ENVIRONMENTAL HAZARDS

EDITED BY SHERMAN HOLLAR

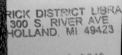

Britannica®
Educational Publishing

IN ASSOCIATION WITH

ROSEN
EDUCATIONAL SERVICES

Published in 2012 by Britannica Educational Publishing
(a trademark of Encyclopædia Britannica, Inc.)
in association with Rosen Educational Services, LLC
29 East 21st Street, New York, NY 10010.

First Edition

Britannica Educational Publishing
Michael I. Levy: Executive Editor, Encyclopædia Britannica
J.E. Luebering: Director, Core Reference Group, Encyclopædia Britannica
Adam Augustyn: Assistant Manager, Encyclopædia Britannica

Anthony L. Green: Editor, Compton's by Britannica
Michael Anderson: Senior Editor, Compton's by Britannica
Sherman Hollar: Associate Editor, Compton's by Britannica

Marilyn L. Barton: Senior Coordinator, Production Control
Steven Bosco: Director, Editorial Technologies
Lisa S. Braucher: Senior Producer and Data Editor
Yvette Charboneau: Senior Copy Editor
Kathy Nakamura: Manager, Media Acquisition

Rosen Educational Services
Alexandra Hanson-Harding: Editor
Nelson Sá: Art Director
Cindy Reiman: Photography Manager
Matthew Cauli: Designer, Cover Design
Introduction by Alexandra Hanson-Harding

Library of Congress Cataloging-in-Publication Data

Poisoning planet Earth : pollution and other environmental hazards / edited by Sherman Hollar.—1st ed.
    p. cm.—(The environment, ours to save)
"In association with Britannica Educational Publishing, Rosen Educational Services."
Includes bibliographical references and index.
ISBN 978-1-61530-508-7 (lib. bdg.)
1. Pollution—Juvenile literature. 2. Hazardous substances—Environmental aspects—Juvenile litera-
ture. I. Hollar, Sherman.
TD176.P645 2012
363.7—dc22

                                                                                    2010051416

Manufactured in the United States of America

On the cover (front inset): Oil covered cormorant swims in polluted waters off Bahrain in the Arabian
Gulf. *Mike Hill/Oxford Scientific/Getty Images*

Front cover background, back cover, page 3: Industrial chimneys emitting smoke. *Shutterstock.com*

Interior background © www.istockphoto.com/Manfredxy

CONTENTS

"It's an ill bird that fouls its own nest."
–English proverb

We live in an age of marvels— ideas, people, and goods speedily fly around the world. Stores are crammed with tempting goods such as cell phones, computers, and exotic foods. But when we buy new things, our planet is paying a price, too. That price can be fragile environments destroyed by overfarming, streams poisoned by factory toxins, and filthy air from fossil fuels spewing from the trucks carrying our goods to market.

Modern humans and their industrial ways are an outsized burden on the planet. One reason is that there are just so many of us. The United Nations predicts that by 2050 almost 9.2 billion people will be living on Earth—almost four times the number of people living on Earth in 1950. It isn't just our numbers, though. We have forgotten the early childhood lesson: If you make a mess, clean it up.

In this volume, you will learn how cities like Los Angeles are frequently shrouded with brown, smoggy skies, caused by the burning of fossil fuels, especially gas emissions from cars. Acid rain—precipitation mixed with gases released from coal-burning plants—kills fish and forests and damages many structures, including important

An aerial photo taken on Oct. 8, 2010, shows a reservoir of red sludge pouring through the walls of a dam at an aluminum plant near Kolontar, Hungary. The ecological disaster began five days earlier when the walls gave way, allowing waves of toxic red mud to sweep through the nearby small village, killing five people. **Attila Kisbenedek/AFP/Getty Images**

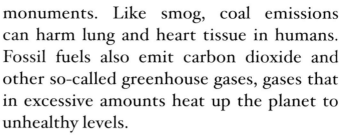

monuments. Like smog, coal emissions can harm lung and heart tissue in humans. Fossil fuels also emit carbon dioxide and other so-called greenhouse gases, gases that in excessive amounts heat up the planet to unhealthy levels.

You will also learn how fertilizers containing chemicals like nitrogen, which plants use for growth, can end up in waterways if farmers have poor soil conservation techniques. This nitrogen can make so much algae blossom that it chokes out all other life in the water.

And then there is the problem of what to do with the ever-increasing mountains of garbage that pile up in landfills. One solution is recycling. The United States recycles about 30 percent of its solid waste, for instance. But it also makes more trash than many other nations; approximately 4.5 pounds (2 kilograms) of garbage per person are generated each day in the United States—significantly more than the 2.5 to 4 pounds (1.1 to 1.8 kilograms) per person that many European nations and Japan make.

Radioactive pollutants have become a major concern since the first atomic bomb was dropped on Hiroshima, Japan, in 1945. Nuclear power plants generate hard-to-store waste, and occasionally disasters have occurred—most notably, the 1986 catastrophe at the Chernobyl nuclear power plant in the Soviet Union. Exposure to radiation can cause cancer and numerous other illnesses, though it may take years for the effects of such exposure to become apparent.

About 620,000 square miles (1.6 million square kilometers) of tropical rainforest are cleared each decade. If deforestation continues at that rate, all tropical forests on Earth—which make up the habitat of two-thirds of the planet's species—will be gone in less than a century. Overpopulation in the Sahel region of Africa has forced hungry people to farm increasingly fragile land bordering the Sahara, contributing to the problem of desertification. In 2006 the UN warned that desertification threatened the livelihoods of about one billion people.

What's next? Individuals can make a difference by consuming less, taking fewer car trips, and recycling. Perhaps most importantly, however, governments, businesses, and environmental organizations need to work together to find large-scale solutions to these problems. To do this will take much political will. But this volume will show why it's so important for us not to foul the nest we rely on for our very lives.

CHAPTER 1
POLLUTION: A GLOBAL CRISIS

Efforts to improve the standard of living for humans—through the control of nature and the development of new products—have also resulted in the pollution, or contamination, of the environment. Much of the world's air, water, and land is now partially poisoned by chemical wastes. Some places have become uninhabitable. This pollution exposes people all around the globe to new risks from disease. Many species of plants and animals have become endangered or are now extinct. As a result of these developments, governments have passed laws to limit or reverse the threat of environmental pollution.

All living things exert some pressure on the environment. Predatory animals, for example, reduce the population of their prey, and animal herds may trample vast stretches of prairie or tundra. The weather could be said to cause pollution when a hurricane deposits tons of silt from flooded rivers into an estuary or bay. These are temporary dislocations that nature balances and accommodates to. Modern economic development, however, sometimes disrupts nature's delicate balance. The extent of environmental pollution caused by humans is already so great that some scientists question whether Earth can continue to support life unless immediate corrective action is taken.

ECOLOGY AND ENVIRONMENTAL DETERIORATION

The branch of science that deals with how living things, including humans, are related to their surroundings is called ecology. Earth supports some 5 million species of plants, animals, and microorganisms. These interact and influence their surroundings, forming a vast network of interrelated environmental systems called ecosystems. The arctic tundra is an ecosystem and so is a Brazilian rainforest. The islands of Hawaii are a relatively isolated ecosystem. If left undisturbed, natural environmental systems tend to achieve balance or stability among the various species of plants and animals. Complex ecosystems are able to compensate for changes caused by weather or intrusions from migrating animals and

A farmer in Vaudreuil, Haiti, works in a cornfield on June 30, 2009.
Thony Belizaire/AFP/Getty Images

are therefore usually said to be more stable than simple ecosystems.

A field of corn has only one dominant species, the corn plant, and is a very simple ecosystem. It is easily destroyed by drought, insects, disease, or overuse. A forest may remain relatively unchanged by weather that would destroy a nearby field of corn, because

These passenger pigeons were painted by the famous American natu-ralist and bird expert John James Audubon (1785-1851) before they became extinct. **Buyenlarge/Archive Photos/Getty Images**

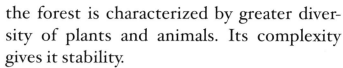

the forest is characterized by greater diversity of plants and animals. Its complexity gives it stability.

Every environmental system has a carrying capacity for an optimum, or most desirable, population of any particular species within it. Sudden changes in the relative population of a particular species can begin a kind of chain reaction among other elements of the ecosystem. For example, eliminating a species of insect by using massive quantities of a chemical pesticide also may eliminate a bird species that depends upon the insect as a source of food.

Such human activities have caused the extinction of a number of plant and animal species. For example, overhunting caused the extinction of the passenger pigeon. The last known survivor of the species died at the Cincinnati Zoo in 1914. Less than a century earlier, the passenger pigeon population had totaled at least 3 billion. Excessive hunting or infringement upon natural habitats is endangering many other species. The great whales, the California condor, and the black-footed ferret are among the endangered animals. Endangered plants include snakeroot, the western lily, and the green pitcher plant.

POPULATION GROWTH AND ENVIRONMENTAL ABUSE

The reduction of the Earth's resources has been closely linked to the rise in human population. For many thousands of years people lived in relative harmony with their surroundings. Population sizes were small, and life-supporting tools were simple. Most of the energy needed for work was provided by the worker and animals. Since about 1650, however, the human population has increased dramatically. The problems of overcrowding multiply as an ever-increasing number of people are added to the world's population each year.

The rate of growth of the world's population has finally begun to slow to slightly more than 1 percent, after reaching an all-time high of more than 2 percent in the early 1960s. In 2007 there were more than 6.6 billion people on the planet. The United Nations predicts that by the year 2050 almost 9.2 billion people will be living on Earth—almost four times the number of people living on Earth in 1950.

The booming human population is concentrated more and more in large urban areas. Many cities now have millions of inhabitants.

15

In less developed countries of Asia, Africa, and Latin America, many of these cities are overpopulated because of an influx of people who have left rural homes in search of employment. Some farmers have been forced off their land by drought and famine.

Environmental pollution has existed since people began to congregate in towns and cities. Ancient Athenians removed their refuse to dumps outside the main part of the city. The Romans dug trenches outside the city to hold garbage and wastes (including human corpses), a practice which may have contributed to outbreaks of viral diseases.

The adverse effects of pollution became more noticeable as cities grew during the Middle Ages.

In Europe, medieval cities passed ordinances against throwing garbage into the streets and canals, but those laws were largely ignored. In 16th-century England, efforts were made to curb the use of coal in order to reduce the amount of smoke in the air—again with little effect.

Medieval cities were often polluted. Here, masons rebuild the city of Jerusalem, after it was taken by French Crusader Godfrey of Bouillon around the year 1100 CE. **Hulton Archive/Getty Images**

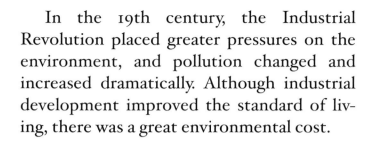
In the 19th century, the Industrial Revolution placed greater pressures on the environment, and pollution changed and increased dramatically. Although industrial development improved the standard of living, there was a great environmental cost.

EFFORTS TO HALT POLLUTION

The solution of some pollution problems requires cooperation at regional, national, and international levels. In the United States laws have been passed to regulate the discharge of pollutants into the environment. The Environmental Protection Agency (EPA), formed in 1970, oversees most federal antipollution activity. The National Environmental Policy Act also mandated the use of environmental impact statements, which require that businesses or governments examine alternatives and acknowledge the possible harmful effects of such activities as opening new factories, building dams, and developing new oil wells. Such international organizations as Greenpeace have become ever more dedicated to preventing environmental abuses and heightening public awareness of environmental issues.

On October 7, 2010, a model holds a recharging cord for the new Chevrolet Volt electric car at the Orange County Auto Show in Anaheim, California. The Volt can travel up to 50 miles on one overnight charge before switching to the gas generator which extends its range to hundreds of miles from one tank of gas. **Mark Ralston/AFP/Getty Images**

The 1970s were a time of great public awareness of the environment. The Clean Air Act, the Safe Drinking Water Act, and the Comprehensive Environmental Response, Compensation, and Liability Act of 1980 (known as Superfund) are among the laws that set standards for healthy air and water and the safe disposal of toxic chemicals. In 1990 President George Bush signed the

THE DEEPWATER DISASTER

One example of the work of the EPA took place as a result of the largest marine oil spill in history. On April 20, 2010, the Deepwater Horizon oil rig—located in the Gulf of Mexico, about 4 miles (66 km) off the coast of Louisiana—exploded. According to the U.S. government, up to 60,0000 barrels of oil a day were spilled after the explosion. The flow of oil was not stopped until September 17.

The petroleum that had leaked from the well before it was sealed formed a slick extending over thousands of square miles of the Gulf of Mexico. Thousands of birds, mammals, and sea turtles were plastered with oil. Birds were particularly vulnerable to its effects, and many perished. Economic prospects in the Gulf Coast states were dire, as the spill impacted many of the industries upon which residents depended. More than a third of federal waters in the gulf were closed to fishing at the peak of the spill, due to fears of contamination.

The various cleanup efforts were coordinated by the National Response Team, a group of government agencies headed by the U.S. Coast Guard and the EPA. BP, Transocean, and several other companies were held liable for the millions of dollars in costs accrued. Following demands by President Barack Obama, BP created a $20 billion compensation fund for those affected by the spill.

Clean Air Act of 1990, the second amending legislation since the original Clean Air Act of 1970. The new law called for reductions in emissions of sulfur dioxide and nitrogen oxide by half, carbon monoxide from vehicles by 70 percent, and other emissions by 20 percent. The number of toxic chemicals monitored by the EPA would increase from 7 to about 250, and industry would be required to control their waste release by means of the best technology available. It also called for promoting alternative fuels. By 2007 the EPA had moved to reduce mercury emissions from power plants and was installing new regulations for exhaust from buses, trucks, and other diesel-powered vehicles.

Internationally, the United Nations Framework Convention on Climate Change adopted the Kyoto Protocol in 2005, a treaty that committed its signatories to develop national programs to reduce their emissions of greenhouse gases. These gases, including carbon dioxide, trap energy from the sun, heating up the Earth. Although the treaty was mired in political debates and most participating countries were not able to meet their goals, the protocol was heralded as a step in the right direction toward an international agreement on environmental policy.

CHAPTER 2
AIR POLLUTION

Factories and transportation depend on huge amounts of fossil fuels. Billions of tons of coal and oil are consumed around the world every year. When these fuels burn they introduce smoke and other, less visible, by-products into the atmosphere. Although wind and rain occasionally wash away the smoke given off by power plants and automobiles, the cumulative effect of air pollution poses a grave threat to humans and the environment.

THE DANGERS OF SMOG

In many places smoke from factories and cars combines with naturally occurring fog to form smog. For centuries, London, England, has been subjected to the danger of smog, long recognized as a potential cause of death, especially for elderly persons and those with severe respiratory ailments. Air pollution in London originally resulted from large-scale use of heating fuels.

A widespread awareness of air pollution dates from about 1950. It was initially

This picture taken on Aug. 6, 2003, shows Paris's Eiffel tower on a smoggy day. Jack Guez/AFP/Getty Images

associated with the Los Angeles area. The Los Angeles Basin is ringed for the most part by high mountains. As air sinks from these mountains it is heated until it accumulates as a warm layer that rises above the cooler air from the Pacific Ocean. This results in a temperature inversion, with the heavier cool air confined to the surface. Pollutants also become trapped at surface levels. Because of air-circulation patterns in the Los Angeles Basin, polluted air merely moves from one part of the basin to another part.

A Chinese worker clears a pile of burning rubbish at a dumpsite in the suburb of Beijing on Feb. 17, 2004. Asia's dangerous pollution levels have triggered an alarming increase in asthma. **Goh Chai Hin/AFP/ Getty Images**

Scientists believe that all cities with populations exceeding 50,000 have some degree of air pollution. Burning garbage in open dumps, which still takes place in some countries, causes air pollution.

Other sources include emissions of sulfur dioxide and other noxious gases by electric power plants that burn high-sulfur coal or oil. Industrial boilers at factories also send large quantities of smoke into the air. The process of making steel and plastic generates large amounts of smoke containing metal dust or microscopic particles of complex and sometimes even deadly chemicals.

The single major cause of air pollution is the internal-combustion engine of automobiles. Gasoline is never completely burned in the engine of a car, just as coal is never completely burned in the furnace of a steel mill. Once they are released into the air, the products of incomplete combustion—particulate matter (soot, ash, and other solids), unburned hydrocarbons, carbon monoxide, sulfur dioxide, various nitrogen oxides, and ozone— undergo a series of chemical reactions in the presence of sunlight. The result is the dense haze characteristic of smog. Smog may appear brownish in color when it contains high concentrations of nitrogen dioxide, or

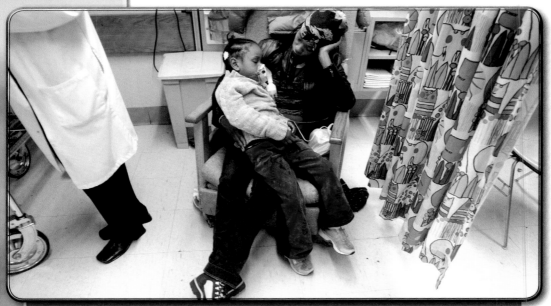

Lizette Samuels holds daughter Starasia Platt, age five, who is being treated for an asthma attack in the emergency room at Coney Island Hospital on Oct. 11, 2002, in the Brooklyn borough of New York City. Mario Tama/Getty Images

it may look blue-grey when it contains large amounts of ozone. In either case, prolonged exposure will damage lung tissue.

Air pollution has enormous consequences for the health and well-being of people worldwide. Contaminants in the air have been implicated in the rising incidence of asthma, bronchitis, and emphysema, a serious and debilitating disease of the lung's air sacs. In addition, current studies suggest that air pollution may be linked to heart disease.

OTHER AIR POLLUTANTS

In the mid-1970s, people became aware of
the phenomenon called acid rain. When fossil fuels such as coal, gasoline, and fuel oils

*A forest pathologist examines pines damaged by ozone in San
Bernardino National Forest, California.* James P. Blair/National
Geographic Image Collection/Getty Images

are burned, they emit sulfur, carbon, and nitrogen oxides into the air. These oxides combine with particles of water in the atmosphere and reach Earth as acid rain, snow, hail, sleet, or fog. A special scale, called the pH scale, measures whether a liquid (including rain and snow) is acidic or basic (alkaline). The pH scale is used to describe the concentration of electrically charged hydrogen atoms in a water solution. The scale rates a substance from 0 to 14. A pH of 7, as in distilled water, means that the solution is neutral. A pH above 7 means the solution is basic; below 7 means the solution is acidic. Normal rainwater has a pH around 5.6.

Although the National Center for Atmospheric Research has recorded storms in the northeastern United States with a pH of 2.1, which is the acidity of lemon juice or vinegar, by the early 21st century the most acidic precipitation in the United States had an average pH of 4.3. In Canada, Scandinavia, and the northeastern United States, acid rain is blamed for the deaths of thousands of lakes and streams. These lakes have absorbed so much acid rain that they can no longer support the algae, plankton, and other aquatic life that provide food and nutrients for fish.

ACID RAIN

During the 20th century acid rain was recognized as a leading threat to Earth's environments. Most acid rain comes from fossil fuel emissions produced in the industrialized Northern Hemisphere—the United States, Canada, Asia, and most of Europe. Acid rain is devastating to all forms of life, but its effects are especially severe in freshwater habitats such as lakes, rivers, and streams. Few aquatic organisms can survive in acidic conditions.

Prevailing winds can carry acidic pollutants around the globe. Research in the 1990s provided strong evidence that emissions from coal-powered electric generating stations in the midwestern United States, such as in Indiana and Ohio, caused the severe acid-rain problem in eastern Canada and the northeastern United States.

Structures made of stone, metal, and cement can be damaged or destroyed by acid rain. Some of the world's great monuments, including the Giant Buddha of Leshan, China, originally carved into the side of a river bank in 713 CE, have deteriorated substantially due to acid rain.

Acid rain can also have serious effects on human health. Particles of sulfur dioxide and nitrogen oxides are easily inhaled and may cause respiratory diseases such as asthma and bronchitis.

Tourists view the 234-foot-tall (71-meter-tall) tall Leshan Giant Buddha, carved into a river bluff in Sichuan Province, China. Some environmental experts warn that worsening pollution in the province has caused acid rain, which in turn is damaging the precious World Heritage– listed artifact. Liu Jin/AFP/Getty Images

By the early 21st century pollutant controls and emission reductions had considerably reduced industrial sulfur dioxide emissions in the United States, Canada, and Europe. Germany and several other countries made substantial progress in recovering damaged forests. However, much of Asia, particularly southwestern China, continued to experience highly acidic precipitation and its associated problems.

Acid rain also damages buildings and monuments, including centuries-old relics such as Rome's Colosseum. Scientists are concerned that the deaths of thousands of trees in the forests of Europe, Canada, and the United States may be the result of acid rain.

Another troubling form of air pollution comes from a variety of human-made chemicals called chlorofluorocarbons, also known as CFCs. These chemicals are used for many industrial purposes, ranging from solvents used to clean computer chips to the refrigerant gases found in air conditioners and refrigerators. CFCs combine with other molecules in Earth's upper atmosphere and then, by attaching themselves to molecules of ozone, transform and destroy the protective ozone layer. The result has been a sharp decline in the amount of ozone in the stratosphere. At ground level, ozone is a threat to our lungs, but in the upper atmosphere ozone works as a shield to protect against ultraviolet radiation from the sun. If the ozone shield gets too thin or disappears, exposure to ultraviolet radiation can cause crop failures and the spread of epidemic diseases, skin cancer, and other disasters.

BHOPAL DISASTER

On Dec. 3, 1984, about 45 tons of the dangerous gas methyl isocyanate escaped from an insecticide plant that was owned by the Indian subsidiary of the American firm Union Carbide in the city of Bhopal, India. The gas drifted over the densely populated neighborhoods around the plant, killing thousands of people immediately and creating a panic as tens of thousands of others attempted to flee. The final death toll was estimated to be between 15,000 and 20,000. Some half a million survivors suffered respiratory problems, eye irritation or blindness, and other maladies resulting from exposure to the toxic gas; many were awarded compensation of a few hundred dollars. Investigations later established that substandard operating and safety procedures at the understaffed plant had led to the catastrophe. In 1998 the former factory site was turned over to the state of Madhya Pradesh.

In the early 21st century more than 400 tons of industrial waste were still present on the site. Neither Dow Chemical Company, which bought out the Union Carbide Corporation in 2001, nor the Indian government had properly cleaned the site. Soil and water contamination in the area was blamed for chronic health problems and high instances of birth defects in the area's inhabitants. In 2004 the Indian Supreme Court ordered the state to supply clean drinking water to the residents of Bhopal because of groundwater contamination. In 2010 several former executives of Union Carbide's India subsidiary — all Indian citizens — were convicted by a Bhopal court of negligence in the disaster.

Two victims sit with patches on their eyes on Dec. 3, 1984, in Bhopal, India. The Union Carbide chemical plant leaked poisonous methyl isocyanate gas into the air, killing thousands. Many others suffered symptoms from exposure to the gas. Pablo Batholomew/Getty Images

In late 1987, more than 20 nations signed the Montreal Protocol to limit the production of CFCs and to work toward their eventual elimination; by 2007 more than 190 countries had joined the agreement. The production of CFCs in developed countries ended in 1996, and now amendments to the pact call for reducing and eliminating the use of hydrochlorofluorocarbons, which replaced CFCs.

Although the release of toxic chemicals into the atmosphere is against the law in most countries, accidents can happen, often with tragic results. In 1984, in Bhopal, India, a pesticide manufacturing plant released a toxic gas into the air that within a few hours caused the deaths of more than 2,000 people.

34

CHAPTER 3
WATER AND LAND POLLUTION

S ince the beginning of civilization, water has been used to carry away unwanted refuse. Rivers, streams, canals, lakes, and oceans are currently used as receptacles for every imaginable kind of pollution. Similarly, many solid and liquid waste materials are deposited on land or underground in a manner that can contaminate the soil and groundwater (water that occurs below the surface of the Earth, where it occupies all or part of the void spaces in soils or rocks), threaten public health, and cause unsightly conditions and nuisances.

CONTAMINATED WATERS

Water has the capacity to break down or dissolve many materials, especially organic compounds, which decompose during prolonged contact with bacteria and enzymes. Waste materials that can eventually decompose in this way are called biodegradable.

Pollution Speeds the Death of a Lake

1. Pollution tinges the cold, oxygen-rich waters of a lake in which trout had been thriving.

⬚ pollutants

⬚ algae

⬚ silt

2. Pollutants dumped into the lake carry plant nutrients (phosphates and nitrates), causing the growth of algae and other water plants. The warmer water now supports bass and perch. Silt from eroded uplands begins to displace the lake water.

3. Algae and vegetation multiply as city sewage and factory wastes pour into the lake. Carnivorous fish are replaced by such plant eaters as bluegills and minnows.

4. In the later stages of a lake's death, algal accumulation becomes dense. Scavengers, such as catfish and carp, feed on decomposed algae and other waste in the silty lake bottom.

5. Pollution finally generate so much plant life that most of the lake's oxygen is consumed in plant decay. Fish cannot live in the oxygenless water. Rooted plants extend further into the lake, continually reducing it. borders. The lake is considered dead and will continue to fill with silt until entirely dry.

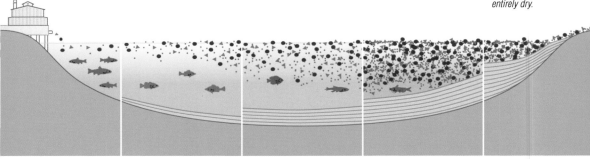

Encyclopædia Britannica, Inc.

They are less of a long-term threat to the environment than are more persistent pollutants such as metals, plastics, and some chlorinated hydrocarbons (chemicals used in some insecticide). These substances remain in the water and can make it poisonous for most forms of life. Even biodegradable pollutants can damage a water supply for long periods of time. As any form of contamination accumulates, life within the water starts to suffer. Lakes are especially vulnerable to

Chinese soldiers remove algae from a beach next to the Olympic Sailing Center in the city of Qingdao on July 5, 2008, before the Beijing Olympics that took place that summer. **Mark Ralston/AFP/ Getty Images**

pollution because they cannot cleanse themselves as rapidly as rivers or oceans.

A common kind of water pollution is the effect caused by heavy concentrations of nitrogen and phosphorus, which are used by plants for growth. The widespread use of agricultural fertilizers and household detergents containing these elements has added large amounts of plant nutrients to many bodies of water. In large quantities, nitrogen and phosphorus cause tiny water algae to bloom, or grow rapidly. When the algae die, oxygen is needed to decompose them. This creates an oxygen deficiency in the water, which causes the death of many aquatic animals. Plant life soon reduces the amount of open water. These events speed up the process of eutrophication, the aging and eventual drying up of a lake.

Sedimentation, the process of allowing solid material such as dirt and sand to enter bodies of water, also pollutes water. Though sometimes sedimentation is natural, at other times, it is the result of poor soil conservation practices. Sediment fills water-supply reservoirs and fouls power turbines and irrigation pumps. It also diminishes the amount of sunlight that can penetrate the water. In the absence of sufficient sunlight, the aquatic plants that normally furnish the water with oxygen fail to grow.

Factories sometimes turn waterways into open sewers by dumping oils, toxic chemicals, and other harmful industrial wastes into them. In mining and oil-drilling operations, corrosive acid wastes are poured into the water. In recent years, municipal waste treatment plants have been built to contend with water contamination. Some towns, however, still foul streams by pouring raw sewage into them. Septic tanks, used where sewers are not available, and large farm lagoons filled with animal waste may also pollute the groundwater and adjacent streams, sometimes with disease-causing organisms. Even the purified effluent (wastewater) from sewage plants can cause water pollution if it contains high concentrations of nitrogen and phosphorus. Farm fertilizers containing chemicals called nitrates can in some regions fill groundwater, making the water unfit to drink. Agricultural runoff containing dangerous pesticides and the oil, grime, and chemicals used to melt ice from city streets also pollute waterways.

THREATS TO THE LAND

In order to sustain the continually growing human population, current agricultural methods are designed to maximize yields from croplands. In many areas, the overuse of land results in the erosion of topsoil (the rich top layer of earth where plants grow). This soil erosion, in turn, causes the over-silting or sedimentation of rivers and streams.

In 1945, a truck sprays DDT on Jones Beach, Long Island (New York) to eliminate mosquitoes, while beachgoers ignore it. **Gamma-Keystone via Getty Images**

One of the most hazardous forms of pollution comes from agricultural pesticides. These chemicals are designed to deter or kill insects, weeds, fungi, or rodents that pose a threat to crops. When airborne pesticides drift with the wind or become absorbed into the fruits and vegetables they are meant to protect, they can become a source of many illnesses, including cancer and birth defects.

Pesticides are often designed to withstand rain, which means they are not always water-soluble, and therefore they may persist in the environment for long periods of time. Some pests have developed a genetic resistance to these chemicals, forcing farmers to increase the amounts or types of pesticide.

The pesticide DDT (dichlorodiphenyltrichloroethane) provides a well-known example of the dangers of introducing synthetic chemical compounds into the environment. Chemically a chlorinated hydrocarbon, DDT was widely used for many years after World War II. At first it was highly regarded because it killed mosquitoes, which in turn reduced the incidence of malaria throughout the world. Then, evidence began to show that DDT might

Only one of twelve mallards hatched from eggs affected by DDT, near Laurel, Maryland. **James P. Blair/National Geographic Image Collection/Getty Images**

be doing more harm than good. DDT, like other chemically stable pesticides, is not readily biodegradable. In addition, many species of insects rapidly develop populations resistant to DDT. The chemical accumulates in insects that then become the diet of other animals, with toxic effects

Recycling Bin

✔ **PAPER:**
Newspaper, magazines, home shopping catalogues, white directories, computer paper, junk mail (remove all plastic wrapping), cardboard, cereal boxes

✔ **METAL:**
Rinsed out food and drink cans

✔ **PLASTIC:**
Rinsed out

✘ **NO** Glass

✘ **NO** Textiles or shoes

✘ **NO** Yellow Pages

✘ **NO** Yoghurt pots, plastic food trays, margarine tubs, plastic bags

✘ **NO** Garden or food waste

A recycling bin waits to be emptied on March 2, 2010, near Leatherhead, England. Peter Macdiarmid/Getty Images

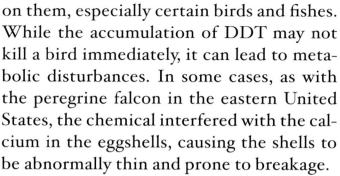

on them, especially certain birds and fishes. While the accumulation of DDT may not kill a bird immediately, it can lead to metabolic disturbances. In some cases, as with the peregrine falcon in the eastern United States, the chemical interfered with the calcium in the eggshells, causing the shells to be abnormally thin and prone to breakage.

Although DDT has been banned in the United States and most other countries, it is still manufactured and used in some parts of the world. Many other pesticides also have been banned. Thousands of pesticides remain in use and, in some cases, their agricultural value may balance out their risks.

Some urban areas are beginning to experience a serious problem regarding the disposal of garbage and hazardous wastes, such as solvents and industrial dyes and inks. In many areas landfill sites are approaching their full capacity and many municipalities are turning to incineration as a solution. Giant high-temperature incinerators have become another source of air pollution, however, because incineration ashes sometimes contain very high concentrations of metals as well as dioxins, a dangerous family of chemical poisons.

RECYCLING

Successful recycling programs depend on several factors. There must be a general awareness of the problems caused by solid-waste disposal and an effective, inexpensive method for separating and collecting the recyclable materials. It also must be economically feasible for industries to use and market recycled materials. In 1976 the United States Congress passed the Resource Conservation and Recovery Act, encouraging states to formulate solid-waste recovery plans. Many states set up special departments to assist local communities in their recycling efforts. Some communities adopted legislation that gives consumers the option of returning containers in exchange for a small deposit paid at the time of purchase.

In the United States, more than 200 million tons of solid waste are generated every year. This amounts to about 4.5 pounds (2 kilograms) per person per day. In metropolitan areas, the daily production of solid waste is usually higher. The overall rate of waste generation is generally lower in Japan and most European countries, ranging from about 2.5 to 4 pounds (1.1 to 1.8 kilograms) produced per person per day.

Recycling efforts usually involve the salvage of materials associated with disposable products—packages, bottles, and labels. The cost of disposing of the solid-waste materials—mainly paper, glass, aluminum, and steel—has steadily increased. In many cases the land used for garbage disposal, known as landfill areas, is too valuable to use as a dumping ground. As existing landfills reach their capacity, many municipalities turn to recycling programs as a relatively inexpensive alternative to landfill disposal.

Bales of paper and cardboard products sit outside the Pratt Industries USA Inc. paper recycling and box manufacturing facility in the Staten Island borough of New York City on, June 28, 2010. Bloomberg via Getty Images

One answer to the garbage problem is recycling. Most towns in the United States encourage or require residents to separate glass and aluminum cans and bottles from other refuse so that these substances can be melted down and reused. According to the Environmental Protection Agency (EPA), the United States recycles more than 30 percent of its garbage. This rate has about doubled over the last 15 years.

The European Union, composed of 27 countries, sets recycling requirements that are met with varying degrees of success. Studies from 2004 show that Greece recycles only 10 percent of its waste while putting 90 percent in landfills. On the opposite end of the spectrum, Denmark recycles about 30 percent, incinerates about 60 percent, and sends a mere 10 percent to landfills.

CHAPTER 4
OTHER TYPES OF POLLUTION

Aside from air, water, and land pollution, there are a number of other types of pollution that present significant dangers to the environment and living things. Modern society is especially concerned about threats posed by radioactive substances and by thermal, or heat, pollution. In addition, noise pollution is a growing problem, particularly in urban areas.

RADIOACTIVE POLLUTANTS

Radioactivity has always been part of the natural environment. An example of natural radioactivity is the cosmic radiation that constantly strikes Earth. This so-called background radiation has little effect on most people. Some scientists are concerned, however, that humans have introduced a considerable amount of additional radiation into the environment.

Since the first atomic bomb was dropped on Hiroshima, Japan, on Aug. 6, 1945, there

The mushroom cloud of the atomic bomb that exploded above Hiroshima, Japan, on August 6, 1945. **Roger Viollet/Getty Images**

has been an increased awareness of the environmental threat posed by nuclear weapons and radioactive fallout. Many scientists are concerned about the long-term environmental impacts of full-scale nuclear war. Some suggest that the large amounts of smoke and dust thrown into the atmosphere during a nuclear explosion would block out the sun's light and heat, causing global temperatures to drop.

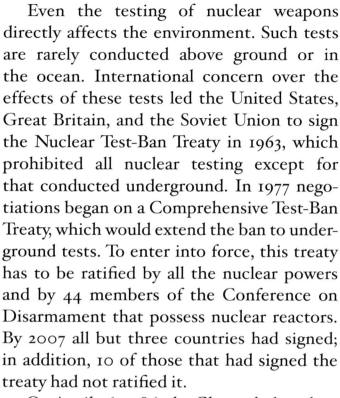

Even the testing of nuclear weapons directly affects the environment. Such tests are rarely conducted above ground or in the ocean. International concern over the effects of these tests led the United States, Great Britain, and the Soviet Union to sign the Nuclear Test-Ban Treaty in 1963, which prohibited all nuclear testing except for that conducted underground. In 1977 negotiations began on a Comprehensive Test-Ban Treaty, which would extend the ban to underground tests. To enter into force, this treaty has to be ratified by all the nuclear powers and by 44 members of the Conference on Disarmament that possess nuclear reactors. By 2007 all but three countries had signed; in addition, 10 of those that had signed the treaty had not ratified it.

On April 26, 1986, the Chernobyl nuclear power plant in the Soviet Union malfunctioned creating the worst peacetime nuclear disaster. Many details of the Chernobyl accident remain undisclosed, but it is known that the radioactive core of the power plant became exposed, and there was a partial meltdown, releasing large amounts of radioactive materials. Because the medical effects of exposure to nuclear radiation can

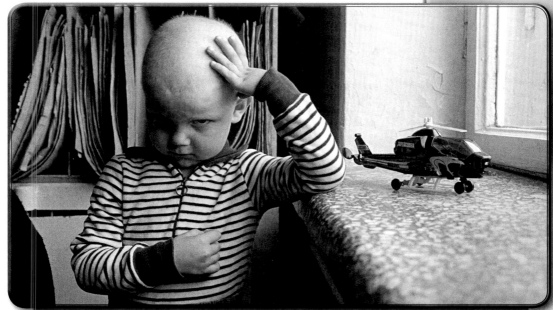

A young cancer patient, Artem Vasyuk, looks shyly toward the camera on August 16, 1996, at the Children's Division of the Oncology Research Institute in Minsk, Belarus. Cancer cases have risen dramatically since the explosion at Chernobyl Nuclear Power Station , which spewed more than 50 tons of radioactive fallout across Belarus.
Ezra Shaw/Getty Images

take years to become apparent, it is not yet known how many additional cases of cancer, birth defects, and skin disease will have been caused by the Chernobyl accident; however, it is estimated that thousands of premature deaths will occur as a direct consequence of nuclear radiation poisoning from Chernobyl.

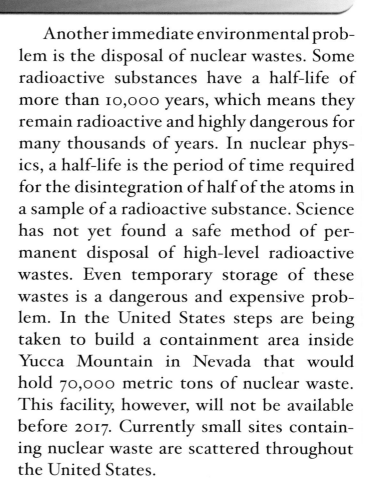

Another immediate environmental problem is the disposal of nuclear wastes. Some radioactive substances have a half-life of more than 10,000 years, which means they remain radioactive and highly dangerous for many thousands of years. In nuclear physics, a half-life is the period of time required for the disintegration of half of the atoms in a sample of a radioactive substance. Science has not yet found a safe method of permanent disposal of high-level radioactive wastes. Even temporary storage of these wastes is a dangerous and expensive problem. In the United States steps are being taken to build a containment area inside Yucca Mountain in Nevada that would hold 70,000 metric tons of nuclear waste. This facility, however, will not be available before 2017. Currently small sites containing nuclear waste are scattered throughout the United States.

THERMAL POLLUTION

While the concept of heat as a pollutant may seem improbable on a cold winter day, at any time of year an increase in water temperature has an effect on water life. Heat can be unnaturally added to streams and lakes

Smoke billows from chimneys at the Chifeng Thermal Power Plant on Feb. 23, 2007, in Chifeng, Mongolia Province, China. **China Photos/ Getty Images**

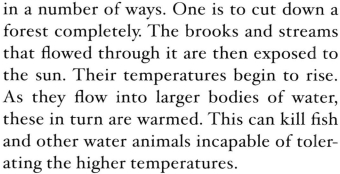

in a number of ways. One is to cut down a forest completely. The brooks and streams that flowed through it are then exposed to the sun. Their temperatures begin to rise. As they flow into larger bodies of water, these in turn are warmed. This can kill fish and other water animals incapable of tolerating the higher temperatures.

Heat pollution is a consequence of the rising energy needs of man. As electric power plants burn fossil fuels or nuclear fuel to provide this energy, they release considerable amounts of heat. Power plants are usually located near bodies of water, which the plants use for heat-dissipation purposes. Some stretches of the Hudson River in New York no longer freeze in winter because of the flow of hot water into the river from adjacent power plants. Living things—especially such cold-blooded animals as fish—are very sensitive to even small changes in the average temperature. Because of the added heat in waters affected by power plants, many aquatic habitats may be undergoing drastic change. In some instances, the warmer water may cause fish eggs to hatch before their natural food supply is available. In other instances, it may prevent fish eggs from hatching at all.

NOISE POLLUTION

The hearing apparatus of living things is sensitive to certain frequency ranges and sound intensities. Sound intensities are measured in decibels. For example, a clap of thunder has an intensity of about 100 decibels. A sound at or above the 120-decibel level is painful and can injure the ear. Likewise, a steady noise at just 75 decibels over numerous hours has the potential to harm hearing. Noise pollution is becoming an unpleasant fact of life in cities, where the combination of sounds from traffic and building construction reverberates among high-rise buildings, creating a constant din.

Jay-Z and Drake perform at Yankee Stadium on Sept. 13, 2010, in New York, New York. Kevin Mazur/ WireImage/Getty Images

In addition, the intense volume at which some popular music, especially heavy metal rock and hip-hop music, is played has resulted in the loss of some or all of the hearing of a few musicians and members of their audiences. There is some evidence that extreme levels of noise can cause stress and produce other harmful effects on human health and on work performance.

Gases such as carbon dioxide, methane, nitrous oxide, and water vapor occur in the environment naturally. These so-called greenhouse gases absorb radiation emitted from Earth's surface and direct it back to Earth, preventing radiant heat loss. This process is known as the greenhouse effect. However, since the Industrial Revolution began to prosper in the 19th century, people have aggressively added more of these gases into the atmosphere through the burning of fossil fuels, the widespread decimation of forests, the raising of large herds of cattle, and other methods. The increase of these gases means that more heat is trapped within Earth's atmosphere, leading to rising global temperatures.

Scientists at the National Aeronautics and Space Administration's Goddard

Institute for Space Studies confirm that the five warmest years worldwide since the late 1880s have all taken place since 1998. Four of the last five have taken place in the 21st century. Scientific data suggests that this trend is likely to continue.

Increased global temperatures have resulted in the steady melting of glaciers and ice caps in the Arctic. Scientific evidence suggests that if polar ice and glaciers continue melting at the current rate of 8 percent per decade, they may disappear completely by 2060. In addition, the melting ice contributes to higher sea levels, currently rising at about 0.08 inch (2 millimeters) per year. If this trend continues, low-lying islands will be completely flooded.

CHAPTER 5
THE DESTRUCTION OF FORESTS

Deforestation is the clearing or thinning of forests, the cause of which is normally implied to be human activity. As such, deforestation represents one of the largest issues in global land use in the early 21st century. Estimates of deforestation traditionally are based on the area of forest cleared for human use, including removal of the trees for wood products and for croplands and grazing lands. In the practice of clear-cutting, all the trees are removed from the land, which completely destroys the forest. In some cases, however, even partial logging and accidental fires thin out the trees enough to change the forest structure dramatically.

ESTIMATES OF WORLDWIDE DEFORESTATION

Conversion of forests to land used for other purposes has a long history. Earth's croplands, which cover about 6 million square miles (15

This undated photo shows workers using elephants for logging in Myanmar. As this Asian nation expands teak logging, more wild elephants are being captured for clear-cutting operations that destroy their habitats, activists say. **AFP/Getty Images**

million square kilometers), are mostly deforested land. More than 4 million square miles (11 million square kilometers) of present-day croplands receive enough rain and are warm enough to have once supported forests of one kind or another. Of these 4 million square miles, only 400,000 square miles (1 million square kilometers) are in areas that would have been cool boreal forests, as in

Leaves display fall colors on Oct. 26, 2007, at the entrance to Woodford State Park in Woodford, Vermont. Fall foliage in the New England region typically reaches its peak color in late October. Most trees in New England are in second-growth forests. **Stan Honda/ AFP/Getty Images**

Scandinavia and northern Canada. Some 800,000 square miles (2 million square kilometers) were once moist tropical forests. The rest were once temperate forests or subtropical forests including forests in eastern North America, western Europe, and eastern China. About another 1.2 million square miles (3 million square kilometers) of forests have been cleared for grazing lands.

Although most of the areas cleared for crops and grazing represent permanent deforestation, deforestation can be transient. About half of eastern North America lay deforested in the 1870s, almost all of it having been deforested at least once since European colonization in the early 1600s. Since the 1870s the region's forest cover has increased, though most of the trees are relatively young. Few places exist in eastern North America that retain stands of uncut old-growth forests.

TROPICAL DEFORESTATION

A wide variety of tropical forests exists. They range from rainforests that are hot and wet year-round to forests that are merely humid and moist, to those in which trees in varying proportions lose their leaves in the dry season, to dry, open woodlands. Because boundaries between these categories are inevitably arbitrary, estimates differ in how much deforestation has occurred in the tropics. Nevertheless, it can be safely said that about 90 percent of the dry forests in the Caribbean, Central America, and the cerrado (savanna and scrub) of Brazil have been cleared. (The drier cerrado lies between the

A worker checks on tapped rubber trees at a rubber plantation in Zongolica, Veracruz state, Mexico, on Thursday, March 4, 2010. **Bloomberg via Getty Images**

humid Amazon Rainforest and the humid forests along the Atlantic coast.) Dry forests in general are easier to deforest and occupy than moist forests and so are particularly targeted by human actions. Worldwide, humid forests once covered an area of about 7 million square miles (18 million square kilometers). Of this, about 4 million square miles (10 million square kilometers) remained in the early 21st century. It is estimated that about 620,000 square miles (1.6 million square kilometers) of tropical forest are cleared each decade. If deforestation continues at that rate, all tropical forests on Earth will be gone in less than a century.

The human activities that contribute to tropical deforestation include commercial logging and land clearing for cattle ranches and plantations of rubber trees, oil palms, and other economically valuable trees. Another major contributor is the practice of slash-and-burn agriculture, or swidden agriculture.

The Amazon Rainforest is the largest remaining block of humid tropical forest, and about two-thirds of it is in Brazil. (The rest lies along that country's borders to the west and to the north.) Detailed studies of Amazon deforestation show that the rate of forest clearing has varied from a low of

SLASH-AND-BURN AGRICULTURE

Slash-and-burn agriculture is a method of cultivation often used by tropical forest root-crop farmers in various parts of the world. Small-scale farmers burn areas of the forest and clear it for planting; the ash provides some fertilization, and the plot is relatively free of weeds. After several years of cultivation, fertility declines and weeds increase. Traditionally, the area was left fallow and reverted to a secondary forest of bush. Cultivation would then shift to a new plot; after about a decade the old site could be reused. By the early 21st century, however, cleared areas were typically maintained in a deforested state permanently. Although traditional practices generally created few greenhouse gases, slash-and-burn techniques are a significant source of carbon emissions when used to initiate permanent deforestation.

about 4,200 square miles (11,000 square kilometers) per year in 1991 to a high of about 12,000 square miles (30,000 square kilometers) per year in 1995. The high figure immediately followed an El Niño, a repeatedly occurring global weather anomaly that causes the Amazon basin to receive relatively little rain and so makes its forests unusually susceptible to fires. Studies in the Amazon also reveal that 4,000–6,000 square miles (10,000–15,000 square kilometers) are partially logged each year, a rate roughly equal to the low end of the forest-clearing estimates cited above. In addition, each year fires burn an area about half as large as the areas that are

The Amazonian rain forest burns as a result of fires started by farmers and ranchers in September 1988 in Rondonia state, Brazil. Stephen Ferry/Getty Images

cleared. Even when the forest is not entirely cleared, what remains is often a patchwork of forests and fields or, in the event of more intensive deforestation, "islands" of forest surrounded by a "sea" of deforested areas.

The effects of forest clearing, selective logging, and fires interact. Selective logging increases the flammability of the forest because it converts a closed, wetter forest into a more open, drier one. This leaves the forest vulnerable to the accidental movement of fires from cleared adjacent agricultural lands and to the killing effects of natural droughts. As fires, logging, and droughts continue, the forest can become progressively more open until all the trees are lost.

Although forests may recover after being cleared, this is not always the case. About 150,000 square miles (400,000 square kilometers) of tropical deforested land exists in the form of steep mountain hillsides. The combination of steep slopes, high rainfall, and the lack of tree roots to bind the soil can lead to disastrous landslides that destroy fields, homes, and human lives. Steep slopes aside, only about one-fourth of the humid forests that have been cleared are exploited as croplands. The rest are abandoned or used for grazing land that often can support only low densities of animals, because the soils underlying much of this land are extremely poor in nutrients.

GLOBAL CONSEQUENCES

Deforestation has important global consequences. Forests sequester (hold on to) carbon in the form of wood and other biomass, or

A father and son help their donkey carry newly-cut trees on their cart into Soro Town in southern Ethiopia, on Aug. 19, 2008. Because wood is the main source of material for both building and fuel, forests in the region are being depleted. **Mike Goldwater/Getty Images**

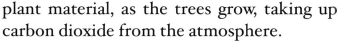

plant material, as the trees grow, taking up carbon dioxide from the atmosphere.

When forests are burned, their carbon is returned to the atmosphere as carbon dioxide, a greenhouse gas that has the potential to alter global climate, and the trees are no longer present to sequester more carbon. In addition, most of the planet's valuable biodiversity is within forests, particularly tropical ones. Moist tropical forests such as the Amazon have the greatest concentrations of animal and plant species of any terrestrial ecosystem. Perhaps two-thirds of Earth's species live only in these forests. As deforestation proceeds, it has the potential to cause the extinction of increasing numbers of these species.

CHAPTER 6

THE PROBLEM OF DESERTIFICATION

Desert environments are expanding in many areas of the world. The spread or encroachment of a desert environment into a nondesert region is a process known as desertification. This process results from a number of factors, including

Because of the lack of food and water in parts of the Sahel, cattle like this bull in northern Niger are dying from starvation. **Issouf Sanogo/AFP/Getty Images**

changes in climate and the influence of human activities.

CLIMATIC FACTORS AND HUMAN ACTIVITIES

Climatic factors in the process of desertification include periods of temporary but severe drought as well as long-term climatic

A man piles up wood for sale on the outskirts of the capital city of N'Djamena, Chad, on Feb. 17, 2009. Authorities had announced a ban on charcoal to help prevent the spread of desertification. **AFP/ Getty Images**

changes toward aridity. For example, in areas where vegetation is already under stress from natural factors, periods of drier than average weather may cause degradation of the vegetation. If the pressures are maintained, soil loss and irreversible change in the ecosystem may ensue, so that areas formerly under savanna or scrubland vegetation are reduced to desert.

There is some evidence that removal of vegetation can also affect climate, causing it to become drier. Bare ground reflects more incoming solar energy and does not heat up as much as ground containing vegetation. Thus, the air that is near the ground does not warm up as much and its vertical movement is reduced, as is atmospheric cooling necessary for condensation and ultimately precipitation to occur.

Human activities often play a major role in desertification. The biological environment of a nondesert region may be degraded by removing vegetation (which can lead to unnaturally high erosion), excessive cultivation, and the exhaustion of surface-water or groundwater supplies for irrigation, industry, or domestic use.

Desertification drains an arid or semiarid land of its life-supporting capabilities.

The process of desertification is extremely difficult to reverse. It is characterized by a declining groundwater table, salinization of topsoil and water, diminution of surface water, increasing erosion, and the disappearance of native vegetation. Areas undergoing desertification may show all of these symptoms, but the existence of only one usually provides sufficient evidence that the process

A worker rests at an iron ore mine on March 31, 2010, in Inner Mongolia Autonomous Region, China. According to state media, Xinghe county has been the main source of sand storms in Beijing due to the expanding desertification. **China Photos/Getty Images**

is taking place. Desertification usually begins in areas made susceptible by drought or over-use by human populations and spreads into arid and semiarid regions.

The main regions currently threatened by desertification are the Sahel region lying to the south of the Sahara desert in Africa, parts of eastern, southern, and northwestern Africa, and large areas of Australia, south-central Asia, and central North America. The arid regions with the longest history of agriculture—from North Africa to China—have generally less well-vegetated deserts.

The present extent of certain of these deserts is thought to be significantly greater than it would be had human impact not occurred. Support for this view is found in various places, such as the several-thousand-year-old rock art from the central Sahara that illustrates cattle and wildlife in regions now unable to support these creatures.

Desertification is not limited to nondes-ert regions. The process can also occur in areas within deserts where the delicate eco-logical balance is disturbed. The Sonoran and Chihuahuan deserts of the American Southwest, for example, have become observably more barren as the wildlife and plant populations have diminished.

THE DESERTIFICATION OF THE SAHEL

Public awareness of desertification increased during the severe drought in the Sahel (1968–73) that accelerated the southward movement of the Sahara. Persistent drought conditions, coupled with substantial growth of both the human and livestock populations in the Sahel, resulted in a gradual desertification of the region.

While climatic variations played a major role in this process, economic and social priorities were involved as well. The introduction of Western technology made it possible for the inhabitants of the Sahel to drill deep water wells. This situation encouraged the herdsmen not only to give up their nomadic way of life and remain near the wells but also to raise more livestock. Overgrazing resulted and, as drought conditions persisted, competition for forage became more and more intense. Herds of goats tore up the remaining indigenous plants by their roots, thereby destroying the ability of the plants to reproduce by themselves. At the same time, the increasing human population led to the cultivating of more and more marginal, ecologically fragile lands for subsistence farming. (The most fertile lands were frequently used to grow cash crops—namely, cotton and peanuts for foreign markets.) Furthermore, Sahelian farmers began reworking the marginal land within one to five years, whereas they had traditionally allowed these lands to remain fallow for 15 to 20 years, giving them ample time to recover. In changing their farming practices, the Sahelians contributed to the destruction of their lands.

As the Sahara moved southward, so did the Sahelians and their livestock. This resulted in further denudation, or stripping of cover, and deforestation. Desertification had set in by the late 1960s, followed by widespread famine. A major international project was undertaken to keep millions of Sahelians alive. Normal rains returned briefly to the Sahel during the mid-1970s; however, since no effective, long-term remedies were applied in animal, agricultural, and human control, outside aid again became necessary in the mid-1980s as drought conditions prevailed once more for a prolonged period (especially in the eastern sections of the sub-Sahara in Ethiopia), causing widespread famine and death. What caused these droughts is still debated, but one of the theories involves

Mauritanian women fetch water on June 8, 2002, in the village of Barkeol, in a garden in the middle of the desert. **Georges Gobet/AFP/ Getty Images**

unusual or anomalous temperature patterns in certain parts of the ocean.

In 1977 the worldwide consequences of desertification were the subject of a UN Conference on Desertification (UNCOD), held in Nairobi, Kenya. In the early 21st century, the UN again highlighted the problem by designating 2006 the International Year of Deserts and Desertification. The UN General Assembly warned that desertification threatened the livelihoods of about one billion people.

76

CONCLUSION

Pollution has accompanied humankind ever since groups of people first congregated and remained for a long time in any one place. As long as there was enough space available for each individual or group, pollution was not a serious problem. With the establishment of permanent settlements by great numbers of people, however, pollution became a significant threat to the environment, and it has remained one ever since.

Through the 19th century, water and air pollution and the accumulation of solid wastes were largely problems of congested urban areas. But with the rapid spread of industrialization and the growth of the human population to unprecedented levels, pollution became a universal problem.

By the later part of the 20th century, an awareness of the need to protect air, water, and land environments from pollution had developed among the general public. Since that time, great efforts have been made to limit the harm to the environment through such methods as air pollution control, banning of destructive pesticides such as DDT,

legal protection of endangered species, cutting back on CFCs to protect the ozone layer, and recycling. In the early 21st century, increasing attention is being paid to reducing the production of greenhouse gases, protecting tropical rainforests worldwide, and implementing soil conservation techniques that may help lessen the consequences of desertification. A deeper understanding of the role of poverty in shaping the crises of deforestation and desertification may help, too. The success of these efforts is far from assured, however. In many cases, significant cooperation between individuals, governments, organizations, and businesses at the international level will be required if meaningful progress is to be achieved.

adjacent Lying next to or near; having a border or point in common.

corrective Serving to correct; having the power of making right, normal, or regular.

corrosive Tending to or having the power to eat away by degrees.

decimation The destruction of a large part of something.

denudation The stripping of all covering; the stripping of land.

deteriorate To make or become worse.

dislocation The act of being put out of place or disrupted.

effluent Liquid (such as sewage of industrial by-products) discharged as waste.

emission Something emitted or discharged.

eutrophication The process by which a body of water becomes enriched in dissolved nutrients, which results in aquatic plant growth, usually causing the depletion of dissolved oxygen.

fallow Describing when land for crops is allowed to lie idle during the growing season.

hydrocarbon A compound containing only carbon and hydrogen.

incineration The burning of something to ashes.

influx A flowing or coming in.

mire To sink or stick fast; to entangle.

noxious Harmful, especially to health.

optimum The best or most favorable amount or degree.

particulate A substance made up of very small separate particles.

refuse Worthless things; trash.

salinization The treating or impregnation of with salt.

signatory A signer with another or others; especially a government bound with others by a signed convention.

turbine An engine that is powered by a series of blades spun around by the pressure of a fluid (as water, steam, or air).

The Canadian Ecology Centre
6905 Hwy 17, P.O. Box 430
Mattawa, ON P0H1V0
Canada
(888) 747-7577
Web site: http://www.canadianecology.ca
The Canadian Ecology Centre offers online
    research resources aimed at conservation
    and development issues and options
    related to the environment as well as
    forestry and mining. Hands-on activities
    and training in outdoors skills are also
    available at the center.

Earth Observatory
Earth Observing System Project Science
    Office
Goddard Space Flight Center
Public Inquiries
Mail Code 130
Greenbelt, MD 20771
Web site: http://earthobservatory.nasa.gov
The Earth Observatory features stories,
    maps, images, and news that emerge
    from NASA satellite missions, in-the-
    field research, and climate models.
    Explore the causes and effects of climatic
    and environmental change through the
    use of this real satellite data.

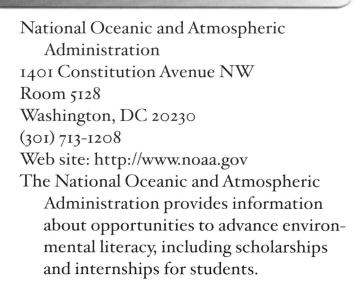

National Oceanic and Atmospheric
   Administration
1401 Constitution Avenue NW
Room 5128
Washington, DC 20230
(301) 713-1208
Web site: http://www.noaa.gov
The National Oceanic and Atmospheric
   Administration provides information
   about opportunities to advance environ-
   mental literacy, including scholarships
   and internships for students.

Natural Resources Canada
580 Booth
Ottawa, ON K1A 0E4
Canada
(613) 995-0947
Web site: http://www.nrcan-rncan.gc.ca/com
Natural Resources Canada offers access to
   its extensive library collection. Posters
   detailing the effects of climate change and
   other issues are available for download.

U.S. Geological Survey
12201 Sunrise Valley Drive
Reston, VA 20192
Web site: http://pubs.usgs.gov/gip
(888) ASK-USGS (275-8747)

The U.S. Geological Survey provides information on ecosystem and environmental health, natural hazards, natural resources, the effects of climate and land-use change, and more. Many USGS data, maps, products, and services are available for the budding scientist interested in deserts, geology, or other related topics.

World Wildlife Fund
1250 Twenty-Fourth Street NW
P.O. Box 97180
Washington, DC 20090-7180
Web site: http://www.worldwildlife.org/climate/index.html
Learn the basics about different environments such as forests and deserts and background on how climate change and other environmental issues are affecting them.

WEB SITES

Due to the changing nature of Internet links, Rosen Educational Services has developed an online list of Web sites related to the subject of this book. This site is updated regularly. Please use this link to access the list:

http://www.rosenlinks.com/teos/pois

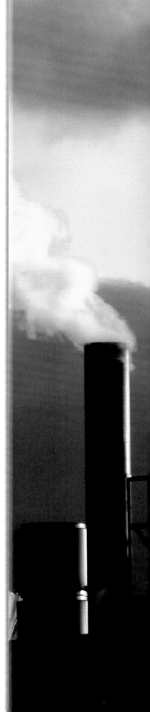

BIBLIOGRAPHY

Bjornlund, Lydia D. *Deforestation* (ReferencePoint Press, 2010).

Bowden, Rob. *Transportation: Our Impact on the Planet* (Raintree, 2004).

Brezina, Corona. *Disappearing Forests: Deforestation, Desertification, and Drought* (Rosen, 2009).

Calhoun, Yael, ed. *Water Pollution* (Chelsea House, 2005).

Carson, Rachel. *Silent Spring*, 40th anniversary ed. (Houghton, 2002).

De Rothschild, David. *The Global Warming Survival Handbook* (Rodale, 2007).

Hill, M.K. *Understanding Environmental Pollution*, 2nd ed. (Cambridge Univ. Press, 2004).

Kidd, J.S. *Air Pollution* (Chelsea House, 2007).

Kusky, Timothy M. *Climate Change: Shifting Glaciers, Deserts, and Climate Belts* (Facts On File, 2009).

Morgan, Sally. *Acid Rain* (Sea to Sea, 2007).

Spilsbury, Richard. *Deforestation Crisis* (Rosen Central, 2010).